DUCK FARMING FOR BEGINNERS

Your Guide to Raising Happy, Healthy Waterfowl Essential Tips and Techniques for New Avian Farmers: Starting and Maintaining a Successful Poultry Farm

BRIYAN GREENWALT

Table of Contents

CHAPTER ONE 9

  Choosing the Right Duck Breeds 9

    Setting Up the Habitat: 9

    Meeting Basic Needs: 10

    Daily Care and Management: 10

    Choosing the Right Duck Breed. ... 11

    Housing & Equipment 12

CHAPTER TWO 13

  Feeding and Nutrition 13

  Health and Disease Management .. 13

  Breeding & Reproduction 14

  Designing A Duck Coop 15

  Suitable Bedding Materials 16

CHAPTER THREE 17

  Providing adequate space. 17

The importance of proper drainage and cleanliness 17

Tips For Predator Proofing 18

Understanding Duck Dietary Needs: 19

Essential Nutrients for Duck Health and Growth: 19

Commercial vs. Homemade Feeds: Pros and Cons: 20

CHAPTER FOUR 21

Feeding Schedules and Portion Control Guidelines................................... 21

Managing Water Sources for Hydration and Cleanliness: 21

Maintaining Duck Health: 21

Common health difficulties in ducks and their symptoms: 22

Recognizing symptoms of sickness and getting veterinarian care 23

Basic first aid for common injuries: 23

Natural Mating vs. Artificial Insemination: 24

Breeding Season Considerations: .25

CHAPTER FIVE 27

Incubation and Hatching 27

Caring for Ducklings: 27

Proper storage and preservation strategies: 29

Adopting sustainable egg production 30

Selling Duck items: 31

CHAPTER SIX 33

Pricing Considerations and Competitive 33

Creating Marketing Materials Effective 33

Building Customer Relationships...34

Navigating Regulations: 34

Liability Insurance and Risk Management: 36

Staying Informed........................36

Predator Management: Strategies for Prevention and Deterrence 37

CHAPTER SEVEN 39

Managing Flock Aggression and Hierarchy Issues. 39

Addressing Dietary Imbalances and Nutritional Deficits.....................39

Managing Parasites and Diseases in Your Flock................................41

CHAPTER EIGHT 43

Common Questions Answered 43

Key Differences Between Duck and Chicken Farming 43

Maintaining Water Quality in Duckponds 44

Alternative uses for duck products ...46

© 2024 [Briyan Greenwalt]. Reserved all rights.

All content in this book cannot be duplicated, shared, or conveyed in any way, including photocopying, recording, or other electronic or mechanical techniques, without the author's prior consent in writing. The only exceptions are short quotes included in reviews and certain other noncommercial uses allowed by copyright laws.

Disclaimer

The author's study, experience, and understanding of livestock management constitute the basis of the material in this book. Concerning the material contained in this book, the author is neither connected to, nor has any affiliation with, any organization, corporation, or person.

The author makes every effort to ensure that the material is accurate and thorough, but any errors or omissions, as well as any results resulting from the use of this information, are not covered by this statement. We strongly advise readers to consult a specialist for advice unique to their situation.

CHAPTER ONE
Choosing the Right Duck Breeds

Successful farming requires careful selection of duck breeds. Beginners should think about things like climate, purpose (meat, eggs, or both), and available resources. Pekin, Khaki Campbell, and Rouen ducks are popular among beginners due to their versatility and productivity.

Setting Up the Habitat: Making a proper habitat for ducks entails giving enough room, shelter, and access to water. Beginners should make sure their duck house is well-ventilated, predator-proof, and contains nesting sites. Additionally, establishing a gated outside area or pond will allow ducks to engage in natural behaviors such as swimming and foraging.

Meeting Basic Needs: Ducks have unique needs for food, water, and healthcare. Beginners should feed their ducks a balanced diet that includes commercial duck feed, greens, and insects. Fresh, clean water should be available at all times, and regular health checks and immunizations are required to avoid sickness.

Daily Care and Management: Routine care and management activities are critical to ducks' health. Beginners should develop a daily schedule that involves feeding, cleaning the living space, and watching for signs of disease or suffering. If you are keeping ducks for eggs, you must gather eggs regularly and keep them clean.

5. Troubleshooting Common Challenges: Despite careful planning, newcomers may face obstacles like as predators, infections, or environmental conditions. Preventive methods such as predator-proof fences, biosecurity, and weather monitoring can help to reduce these hazards.

In addition, obtaining guidance from experienced duck farmers or participating in internet forums can provide useful information and assistance.

Choosing the Right Duck Breed.

When beginning out in duck farming, choosing the right breed is critical. Consider aspects such as climate compatibility, farming objectives (eggs, meat, or both), and available area. Popular breeds for beginners include the Pekin, Khaki Campbell, and Indian

Runner. Each breed has its own temperament, egg-laying capacity, and meat quality, so consider these factors based on your preferences and requirements. To get your farming endeavor off to a good start, make sure to buy ducklings or eggs from trusted providers.

Housing & Equipment

Creating appropriate housing and getting the necessary equipment are critical elements in duck farming. Build or purchase a secure shelter that protects you from predators and inclement weather.

Ensure that your ducks have ample airflow and space to roam about comfortably. Nesting boxes for egg-laying ducks should be placed in the housing, together with bedding for warmth and cleanliness. In addition, invest in duck-specific feeding and watering equipment to ensure clean water and a balanced diet.

CHAPTER TWO
Feeding and Nutrition

Your ducks' health and production depend on proper feeding. Provide a balanced diet consisting of commercial duck feed supplemented with fresh greens, grains, and protein sources such as insects or mealworms. Adjust feeding amounts based on age, breed, and seasonal changes in nutritional requirements. Ensure that ducks always have access to clean water, since they require it for digestion and overall health.

Health and Disease Management

Maintaining excellent health and preventing infections is critical for successful duck farming. Establish a regular health-monitoring regimen, looking for indicators of disease or injury. Vaccinate ducks against common illnesses and parasites on a suggested schedule. Maintain proper hygiene by keeping your housing and equipment clean, periodically eliminating droppings, and providing clean bedding. Maintain a relationship with a veterinarian who specializes in poultry care for guidance and support as needed.

Breeding & Reproduction

Understanding the breeding and reproduction processes is critical for managing a duck flock. Provide optimal mating conditions, providing a balanced ratio of drakes to ducks.

Monitor mating behavior and egg production to determine viable eggs for hatching. Consider using natural incubation or investing in an incubator to hatch ducklings. Ducklings require appropriate care, including warmth, protection, and access to food and water. Keep accurate records of breeding operations and hatch rates to assess the long-term success of your breeding effort.

Designing A Duck Coop

When building a duck coop, consider its size, ventilation, and security. Ducks require plenty of areas to walk around freely, so make sure your coop is large enough for them to wander.

Adequate ventilation is critical to preventing the accumulation of moisture and ammonia, which can cause respiratory problems. Ensure that adequate security measures are in place to safeguard your ducks from predators.

This includes strong fencing, tight locks, and maybe installing motion-activated lights or alarms.

Suitable Bedding Materials

Choosing the appropriate bedding materials for your duck enclosure is critical to their comfort and health. Choose materials such as straw, hay, or wood shavings, which provide insulation and absorb moisture well.

Avoid using cedar shavings since they might be poisonous to ducks. Keep the bedding clean and dry to minimize bacteria and mold growth, which can lead to respiratory issues and illnesses in ducks.

CHAPTER THREE

Providing adequate space.

Ducks require space not simply to move around, but also to perform natural behaviors such as nesting, roosting, and foraging. Ensure that your duck habitat has ample space for these activities. Provide nesting boxes or locations with acceptable nesting materials, such as straw or hay. Install roosting perches or platforms so ducks may rest comfortably off the ground. Allow access to outdoor spaces for feeding and exercise while keeping them safe from predators.

The importance of proper drainage and cleanliness

Maintaining appropriate drainage and cleanliness in your duck habitat is critical to avoiding health problems and unpleasant odors.

Ensure that the coop floor is gently slanted to help water to drain easily. To keep the surroundings clean and dry, replace used bedding regularly. To prevent contamination and illness spread, clean your water and food dishes daily. Inspect the coop regularly for symptoms of mold, mildew, or pests, and take action to address them as soon as possible.

Tips For Predator Proofing

Predators constitute a big threat to ducks, so you must take precautions to predator-proof your duck habitat. To keep predators out, install robust fencing with minimal openings.

Consider installing electric fencing to provide additional protection, particularly in areas with heavy predator activity. Secure apertures and entrances with locks or latches that predators can't easily enter. Remove any potential hiding places around the coop, such as dense grass or vegetation. Consider utilizing deterrents, like as motion-activated lights or alarms, to scare away prospective predators. Inspect the perimeter regularly for evidence of digging or tampering, and repair any damage as soon as possible.

Understanding Duck Dietary Needs: Ducks require a well-balanced diet to flourish. Their nutritional requirements include proteins, vitamins, minerals, and carbs. Understanding these demands is critical to their health and development.

Essential Nutrients for Duck Health and Growth: Ducks require a variety of nutrients for good health and growth. Proteins are essential for muscle development, whereas vitamins and minerals promote overall health. Carbohydrates supply energy for daily tasks.

A balanced diet for ducks includes grains, vegetables, and protein sources. Grains like corn and wheat supply carbohydrates, whereas greens like lettuce and spinach provide important vitamins. Protein sources such as insects, fish meal, and soybean meal promote healthy muscle development.

Commercial vs. Homemade Feeds: Pros and Cons: Commercial feeds provide convenience and adequate nutrition, whilst homemade feeds allow for customization and may be less expensive. When deciding whether to use commercial or homemade feeds, consider availability, cost, and nutritional content.

CHAPTER FOUR

Feeding Schedules and Portion Control Guidelines

Using a feeding plan ensures that ducks get enough nourishment without overeating. Portion control is critical to avoiding obesity and nutritional imbalances. Provide fresh feed daily, and monitor intake to alter portion amounts accordingly.

Managing Water Sources for Hydration and Cleanliness: Ducks require clean water for hydration and cleanliness. To avoid pollution and disease transmission, ensure that water sources are cleansed and refilled regularly. Proper water management is critical for duck health and welfare.

Maintaining Duck Health: Regular health monitoring is essential for ensuring that your ducks thrive. Keep a look out for symptoms of common health problems, including respiratory troubles, lameness, or unusual behavior. Early detection is critical, therefore implement a regular health check program to properly monitor their condition.

Common health difficulties in ducks and their symptoms: Familiarize yourself with the most common health concerns that ducks confront, such as respiratory infections, botulism, and parasites. Keep an eye out for signs such as coughing, sneezing, decreased appetite, or drooping. Recognizing these signals early allows for rapid intervention, reducing the risk of consequences.

Preventive procedures include vaccines, biosecurity, and hygiene. Prevention is preferable to cure. Vaccinate your ducks against common infections such as duck viral enteritis and avian flu. Implement strong biosecurity measures to avoid pathogen introduction and spread. To reduce the danger of disease, keep their living spaces clean and disinfected regularly.

Recognizing symptoms of sickness and getting veterinarian care: Understand when to seek professional assistance. If you see any chronic or severe symptoms in your ducks, contact a veterinarian immediately. Early intervention can reduce the spread of sickness and increase the likelihood of successful treatment.

Basic first aid for common injuries:
Be ready to provide rapid care for cuts, wounds, and leg injuries. Clean the

wound with an antiseptic solution, then apply a proper bandage and create a pleasant, quiet environment for recuperation. Familiarize yourself with basic first aid skills so you can treat small injuries quickly and effectively.

Duck breeding basics include studying their reproductive anatomy and behavior. Female ducks have a distinct reproductive system, which includes a single oviduct where eggs are produced. Males have a phallus for mating. Understanding these characteristics is critical for effective breeding.

Natural Mating vs. Artificial Insemination: In terms of duck breeding, both natural mating and artificial insemination offer advantages and disadvantages. Natural mating is simpler and more natural, yet it may result in lower fertilization rates. Artificial insemination provides greater control over genetics and breeding results, but it necessitates specialized equipment and experience.

Breeding Season Considerations: Duck breeding is frequently seasonal, driven by elements such as daylight length and temperature. Ducks often breed in the spring and summer, when conditions are favorable. Breeding operations should be planned around these seasons to maximize success rates.

CHAPTER FIVE
Incubation and Hatching

After eggs are laid, they require adequate incubation before hatching. Duck breeders can select both DIY methods, such as employing a broody duck or an incubator, and commercial ones. Each method has advantages and disadvantages, therefore it is critical to choose the most appropriate one based on resources and skills.

Caring for Ducklings: Once ducklings hatch, they require close supervision during the brooding period. Providing a warm and comfortable habitat, adequate nutrition, and access to water are critical for their early development.

Monitoring their health and resolving any concerns as soon as possible is critical to ensuring that they mature into healthy adult ducks.

Maximizing Egg Yield In duck farming, it is critical to consider aspects like as food, lighting, and stress management. Ducks require a well-balanced, nutrient-rich diet for optimal egg production. Ensure they have access to clean water and nutritious food.

To increase egg laying, proper illumination should simulate natural daylight. Reduced stresses, like as overcrowding or abrupt environmental changes, can also enhance egg production.

Collecting and handling fresh eggs carefully is critical for preserving their quality. Check nests regularly, ideally twice a day, to ensure that eggs are collected promptly. Handle eggs cautiously to avoid cracking or damage. Clean soiled eggs using a dry cloth or brush; avoid washing with water, which can dissolve the protective cuticle. To avoid spoiling, keep collected eggs cool, dry, and well-ventilated.

Proper storage and preservation strategies: are essential for keeping eggs fresh. To help keep eggs fresh, Place them with the pointed end down. Keep them refrigerated at temperatures around 45°F/7°C. Eggs can absorb flavors, so avoid storing them near meals with strong odors. If you want to keep eggs for an extended amount of time, consider freezing or pickling them.

Troubleshooting Common Egg-Laying Issues It is critical to quickly identify and resolve common egg-laying issues. Inadequate nutrition, illnesses, and stress can all contribute to issues such as soft-shelled eggs or decreased egg production. Monitor duck health closely, alter their nutrition as needed, and provide a pleasant environment. A veterinarian can assist diagnose and treat any underlying health conditions that are impeding egg production.

Adopting sustainable egg production processes benefits the environment as well as company sustainability. Rotational grazing, organic feed, and good waste management can all help to lessen the environmental impact.

Consider adding solar-powered lighting for increased energy efficiency. Prioritize ethical treatment of ducks by enabling them access to open places and natural activities. Integrating sustainability into egg farming helps ensure long-term viability while addressing consumer demand for ethically produced eggs.

Selling Duck items: To effectively sell your duck items, you must first define your target markets. These could include local customers looking for fresh and sustainable options, restaurants looking for high-quality ingredients, and specialty businesses catering to certain markets. Understanding your audience allows you to personalize your products and marketing activities to their requirements and tastes, boosting the possibility of sales.

Creating a Branding Plan Crafting a good branding plan for your duck products is critical to distinguishing out in the market. This entails developing a distinct identity that resonates with your target audience while emphasizing the distinctive aspects of your products, such as freshness and sustainability. Effective branding allows you to differentiate yourself from the competition and establish a devoted customer base.

CHAPTER SIX
Pricing Considerations and Competitive

Analysis Choosing the appropriate price for your duck products necessitates careful analysis of manufacturing costs, market demand, and rival pricing.

Conducting a competitive study can help you understand industry pricing patterns and uncover possibilities to market your items more competitively. By establishing the perfect mix between affordability and value, you may increase sales while remaining profitable.

Creating Marketing Materials Effective marketing materials are essential for showcasing your duck products and attracting clients. This could include creating eye-catching labels that highlight key product attributes, distributing fliers in local communities and farmers' markets, by utilizing many marketing platforms, you can reach a larger audience and increase interest in your product.

Building Customer Relationships Establishing good customer relationships is essential for sustaining loyalty and repeat business. This includes delivering excellent customer service, swiftly responding to any issues or requests, and asking for feedback to continuously improve your products and services. By focusing on customer satisfaction and engagement, you may build a loyal client base that will support your duck farming business for years to come.

Navigating Regulations: Before getting into duck farming, it's important to grasp the regulatory landscape. This entails obtaining the appropriate permits and licenses for your business. Different regions may have different requirements, thus extensive research and communication with local authorities are required to ensure compliance. Additionally, remaining up to date on animal welfare standards and environmental legislation is critical to maintaining ethical farming practices and minimizing negative environmental impacts.

Zoning and Land Use Considerations: Zoning restrictions play an important part in selecting where you can set up your duck farm. Understanding these requirements guarantees that you select an appropriate place by zoning laws. To avoid conflicts and maximize agricultural efficiency, consider factors such as closeness to residential areas, water sources, and infrastructure. Conducting careful research and speaking with local zoning officials can assist you in making informed land-use decisions.

Liability Insurance and Risk Management: Duck farming, like any other industry, involves inherent hazards. Obtaining liability insurance is critical for protecting your farm from unexpected events like accidents or property damage. Assessing potential risks and applying risk management measures can help to reduce liabilities and protect your investment. To provide comprehensive protection, consult with insurance pros to tailor coverage to your farm's specific needs.

Staying Informed

Duck farming regulations are subject to change, therefore it is critical to stay up to date on new laws and industry standards. Using resources such as industry journals, government websites, and professional organizations can help you stay up to date on regulatory changes and best practices. Networking with other farmers and attending appropriate workshops or seminars can also provide useful insights and assistance in handling legal and regulatory issues effectively.

Predator Management: Strategies for Prevention and Deterrence

When it comes to protecting your duck herd against predators, prevention is essential. Begin by strengthening your farm with a strong fence, which includes subsurface barriers to repel burrowing predators. Employ guardian animals such as dogs or geese to deter potential threats. Consider installing motion-activated lights or sound devices to deter nighttime predators. To keep your ducks safe, evaluate your perimeter for weak points regularly and swiftly patch any breaches that occur.

CHAPTER SEVEN

Managing Flock Aggression and Hierarchy Issues.

Maintaining harmony within your duck flock is critical to their well-being. To manage hostility and build a harmonious order, give your ducks plenty of room to move and forage.

To prevent territorial conflicts, introduce new ducks gradually. Provide an appropriate amount of food and water to reduce competition among flock members. If aggressive behavior persists, separate aggressive ducks temporarily or consider rehoming particularly disruptive birds to create a harmonious environment.

Addressing Dietary Imbalances and Nutritional Deficits

Providing your ducks with a balanced diet is critical for their health and production. Begin by providing a commercial duck feed tailored to fulfill their nutritional requirements. Supplement their diet with fresh greens, cereals, and protein sources such as mealworms or shrimp. Monitor their condition for indicators of malnutrition, such as poor feather quality or lethargy, and change their diet as needed. If you're not sure how to suit your ducks' special dietary needs, consult with a poultry nutritionist.

Managing Parasites and Diseases in Your Flock

Preventing and controlling parasites and illnesses is critical for keeping a healthy duck population. To prevent pathogen spread, keep your coop and nesting spaces clean and disinfected regularly.

To control parasites such as mites and worms, apply poultry-safe insecticides and dewormers exactly as prescribed. Vaccinate your ducks against common infections such as duck viral enteritis and avian influenza to strengthen their immune systems.

Monitor your flock for signs of sickness, like as decreased egg production or respiratory issues, and contact a veterinarian right once if you suspect a problem.

Handling weather-related challenges, such as heat stress, cold exposure, and extreme weather.

Weather extremes can provide substantial obstacles for duck farming, but with appropriate planning, you can reduce their influence on your flock. To avoid heat stress in hot weather, provide plenty of shade and access to cool water.

Install misters or fans in your coop to assist reduce the temperature. In colder climates, insulate your duck coop and give bedding to keep them comfortable. To keep your water from freezing throughout the winter, consider utilizing heat lights or heated waterers

CHAPTER EIGHT

Common Questions Answered

What number of ducks should I start with?

Start with a modest amount, such as 4 to 6 ducks, to get the feel of duck farming without overwhelming yourself. This allows you to learn about their actions and demands without overwhelming yourself. As you acquire experience, you can gradually grow the size of your flock based on your capabilities and objectives.

Key Differences Between Duck and Chicken Farming

Duck and chicken farming differ in a variety of ways. Ducks are typically more cold-tolerant and disease-resistant than chickens, making them suited for a variety of climates and habitats. Ducks and chickens have differing food requirements, with ducks preferring more greens and insects. Ducks lay fewer but larger eggs, with a thicker shells. Understanding these distinctions will allow you to adjust your agricultural operations to your preferred fowl.

Maintaining Water Quality in Duckponds

Maintain water quality in duck ponds by implementing regular cleaning schedules and adequate filtration systems. Remove any debris or algae accumulation regularly, and maintain proper circulation to avoid stagnation. Monitoring water pH levels and adding helpful microorganisms can also help your ducks' water stay clean and healthy. Regular water quality inspections are required to ensure that your ducks have access to clean water, which is critical for their health and well-being.

Ducks living alongside other farm animals

Ducks may coexist peacefully with other farm animals, but proper introductions and supervision are essential. Ensure that other animals, such as hens or goats, do not hurt the ducks, and offer separate feeding sites to avoid competition. Ducks can also assist control pests and weeds in fields when combined with other animals. Monitoring their interactions and providing proper shelter and space can help to ensure harmonious coexistence on your farm.

Alternative uses for duck products

Aside from food, duck products have a variety of other uses. Duck feathers are suitable for making cushions, blankets, and insulating materials. Duck fat is highly valued for cooking, and it can be converted into gourmet cooking oil or used to manufacture soap and candles. Additionally, duck dung is a great fertilizer for gardens and crops, providing a long-term strategy to improve soil health. Exploring these additional uses can help duck farmers reap more rewards than just meat and eggs.

In conclusion, duck farming is a viable investment for beginners looking for a long-term and profitable agricultural endeavor. Throughout this journey, we've looked at everything from duck breed selection and housing to dietary demands and healthcare requirements.

Duck farming has various advantages, including minimal start-up costs, excellent feed conversion rates, and a wide range of market options. Furthermore, ducks are hardy birds that can adapt to a variety of climates and habitats, making them an excellent choice for new farmers.

Furthermore, ducks are more versatile than only meat and eggs. Their feathers, down, and even dung can be used to generate additional revenue or implement ecological practices like composting. This multimodal method increases the economic sustainability of duck farming for beginners.

However, success in duck farming requires painstaking planning, rigorous management, and a thorough understanding of the birds' behavior and demands. Beginners must focus on learning about adequate diet, disease prevention, and optimal housing conditions to ensure their flock's health and well-being.

Furthermore, developing strong networks within the farming community, finding mentorship, and staying current on industry developments are critical steps toward establishing a successful duck farming operation.

To summarize, while duck farming for beginners has its hurdles, the potential for profit and personal fulfillment is significant. With effort, perseverance, and a commitment to constant learning, prospective duck farmers may embark on a journey that not only supports their lives but also adds to the larger agricultural environment. So, if you're thinking about starting duck farming, seize the opportunity, embrace the trip, and see your efforts take off in this thriving sector.

www.ingramcontent.com/pod-product-compliance
Lightning Source LLC
Chambersburg PA
CBHW072020230526
45479CB00008B/307